A Controversial Approach To Music Theory & Musical Scansion

Philip Martorella

A Controversial Approach to Music Theory and Musical Scansion. Copyright © 2018 Philip Martorella. Produced and printed by Stillwater River Publications. All rights reserved. Written and produced in the United States of America. This book may not be reproduced or sold in any form without the expressed, written permission of the authors and publisher.

Visit our website at www.StillwaterPress.com for more information.

First Stillwater River Publications Edition

Library of Congress Control Number: 2018943309

ISBN-13: 978-1-946-30058-4
ISBN-10: 1-946-30058-6

1 2 3 4 5 6 7 8 9 10

Written by Philip Martorella
Cover design by Nathanael Vinbury
Published by Stillwater River Publications, Pawtucket, RI, USA.

The views and opinions expressed in this book are solely those of the author and do not necessarily reflect the views and opinions of the publisher.

Table of Contents

Preface i

Chapter One - The Labels of Music Theory 1

Chapter Two - Analysis of Movement I
of Symphony No. 40 in G minor, measures 1-42,
K. 550 by W.A. Mozart 7

Chapter Three - Analysis of movement I, measures 1-66
from Symphony No. 5 in C minor, Op. 67
by L. van Beethoven 13

Chapter Four - Patschen with Mixed Rhythms 21

Chapter Five - Musical Scansion 25

Chapter Six - Examples of Musical Scansion 29

    Example 1. O Fortuna from Carmina Burana
    by Carl Orff 29

    Example 2: Polonaise-Fantaisie in Ab Major,
    Op. 61 by Frederic Chopin 31

    Example 3. Variations on a Theme of Robert
    Schumann, Op. 20 by Clara Schumann 33

    Example 4: Prelude from Tristan and Isolde
    by Richard Wagner 35

    Example 5: Impromptu in Bb Major, Op. 142, No.3,
    measures 1-18 (the measure numbers are not
    including the repeat) by Franz Schubert 37

    Example 6: Verklärte Nacht, Op. 4 (Transfigured
    Night) by Arnold Schönberg 40

    Example 7: Opening of the First Movement;
    Opening of the Second Movement from the
    Symphony in E minor, Op. 27 by S. Rachmaninoff
    41

Example 8: Danse Russe (Trepak) from
The Nutcracker Suite, Op. 71a, measures 1-7
by Pyotr Tchaikovsky 43

Chapter Seven - An interview with the author
Philip Martorella, pianist, organist, composer,
professor 45

Appendix A 55

Scores Referenced in Chapter Six 67

Appendix B 67

    Teacher-pupil Lineage Tree. 67

Dedication

For Sal, my father and for Jean, my mother.

Acknowledgements

Thank you to John Faiola.

Preface

Before one is to understand the concepts of music theory, one must know what theory itself means. Theory is an amalgamation of ideas or concepts. Musical theory is also an amalgamation of ideas blended or combined together through analytical processes in accordance with visual and audial activity.

Discovery of various elements of music such as pitch, rhythm, basic notations of pitches, degrees of sound, intervallic relationships, chordal structures, and the mathematical calculations of the relationships between rhythm and sound form the basis for theoretical analyses that may exceed far beyond the expectations of the average listener. The average composer would instinctively desire his or her music to be notated on manuscript paper upon a natural inspiration and somehow react to that inspiration before even thinking about what chord or chords are involved, which become unfolded during the process of creation after the initial theme is first and foremost created. Inventions or discoveries were of human origin and never cease to become modified throughout the years. The inventors or discovers of the past are indeed well respected; and, of course, there are always considerable improvements upon their accomplishments by new discoveries, inventions, or modifications. Essentially, it all boils down to new and unprecedented ideas or concepts, theories or practices. Composer Igor Stravinsky decided to "invent" his own scale by omitting the fourth scale degree.

This book is controversial in that it deviates from the traditional norms of theory and the ordinary expectations of structural analyses. One may disagree with some of the ideas that are introduced and that is perfectly acceptable and respectable. I do believe that there is a difference between theory and practice anyway, regardless of any differences of opinion. Some believe that theory is abstract and practice is more concrete. If that is true, then one should think twice about attempting to "reinvent the wheel" too.

Theory is nevertheless a collection of ideas, whereas practice is in opposition to theory, in that it is more about applying oneself by executing a certain talent or implementing an idea to full effect if possible. Therefore, why has the phrase "music theory" been utilized throughout the generations? The answer seems to be about the inclusion of the observations and results of the ideas that have transitioned throughout history and is, of course, ongoing through our own generation. There is the study of traditional harmony along with traditional melody also and, whether it is consonant or dissonant makes no difference. Consonance is thought of as agreeable sonorities of a pleasant, harmonious nature as opposed to dissonance, which is thought of as disagreeable sonorities of an unpleasant, harmonious nature. Regardless of whether it involves tones of conformity or tones of non-conformity, it is much like life itself with all its pleasantries, successes and compatibilities; or, on the other hand, its unpleasantries, struggles and conflicts. One must deal with these aspects of life as they exist, whether we approve or disapprove of these facts. A consonant sonority may be like a breath of fresh air. A dissonant sonority may be a disturbance of some form, whether it be jarring or annoying. Personally, I find that dissonance tends to form a sense of tension, which is abundantly utilized and necessary in music, as consonance tends to form a soothing sense of relaxation or resolution and is also abundantly utilized and necessary. These two components of music, namely, consonance and dissonance, are the basis for our great works of music and are much required and desired to create the necessary elements of musical knowledge, appreciation, and the resultant effects, whether they be dramatic, intense, or soothing and pacifying.

If our current generation consists of what is termed "contemporary, or modern music," then what will the music of our future be termed? If it is a fusion or intermixture of various genres or styles of music, I could envision that for the future perhaps as a possibility. Yet, it would be ultra-modern and maybe a bit more chordally extraneous in various degrees.

Philip Martorella

Chapter One

The Labels of Music Theory

Herein, again, is going to be some controversy. I hereby welcome any refutations regarding my ideas.

I will use mostly the upper case roman numerals for now so as to indicate the labels for each chord. Let's not concern ourselves for now about what is Major and what is minor or diminished or augmented for that matter. Traditionally, the chords are: I This is labeled as the tonic. No problem. II is the supertonic. Sure, it's above (super) the tonic. No problem. III is the mediant. No problem. IV is the subdominant. No problem. V is the dominant. No problem. VI is the submediant? Problem. Yes, it is often thought of not only as the sixth scale degree of a musical scale that is stepwise in its order; but, it is also thought of as being in the middle of the upper tonic and subdominant. This may seem rebellious of me; but, I have a fascination with this inherent desire to modify the label of this chord. It seems senseless to think that the term submediant could also imply under (sub) the mediant, which is III.

I prefer to label it as the superdominant. Therefore, I will. I actually have seen this term once before so it had already been utilized. Continuing on, VII is the leading tone. Problem. I am relabeling this one, too. Historically, the subtonic, as I prefer to label it, was considered as the leading tone triad. The subtonic, after approximately the second quarter of the 20th century included the flatted seventh. Let's revert back to the original way that it was constructed. Specifically, one theorist apparently thought to flatten the seventh after the second quarter of the 20th century and alter the so-called definition of the leading tone triad. I don't like the phrase "leading tone" because it sounds like one is relishing the tendency of the seventh tone itself, whether flat or natural, to intentionally become resolved upward, whether it be a half step or whole step ascending. The ascending movement to resolve is nevertheless plausible by itself. According to music theorist Heinrich Schenker of the 19th-20th centuries, another example of a so-called descending leading tone is not the seventh but the second scale degree as a descending leading tone. By the way, the subtonic is considered "the seventh note of the diatonic scale," according to the thesaurus. Not specific enough? Please feel free to be the judge if you so desire.

All right, now that we have established some bit of controversy about some labels, the following term is another one: Tonicization. This is the tonic chord which supplants or substitutes another chord, creating a temporary replacement of that chord so as to make analysis of a particular section function as if it was in the tonic or "home" key. I have utilized tonicization on various occasions before. Nevertheless, I often wondered why I was dealing with this concept of temporarily changing the key, usually the dominant, although it could have been any other chord other than the actual tonic "home" key. I thought, why not just use the regular structural chords as they are so as to avoid tonicization? After all, tonicization is only a temporary phenomenon. It is not to be confused with modulation, which emerges into a new key anyway by itself and often cadences at least once if not more.

To figure the chords out regularly as they appear, one would need to search the transformational chord structures along with its necessary linear embellishments overarching the main chord structures: whether such embellishments are neighbor note or passing note motions, these would only enhance the basic, fundamental, or implied chord structures to be searched. The chords can be found even in their state of various permutations, configuring into secondary, tertiary, quaternary, quinary, senary, septenary, octonary, nonary, and denary chordal conversions or transformations. To make an interesting analogy of these events, I think about the bloodlines of families and their existing relatives or, to really go off on a tangent, follow the ancestry of such familial bloodlines and most importantly how they are all related, whether close or distant, with regard to relationships. There is a kind of metamorphosis that is experienced. Therefore, with reference to the chordal events, much like the familial bloodlines, the various permutations would then be in order; for example, a secondary dominant of a dominant relates to the tonic in question. If I intended to go further, I could analyze the chord structures, whether vertically rendered or in a rather condensed format regarding the actual or implied chord, even if it is conceived through a horizontal perspective. Moreover, a chord could then be configured as it could be somewhat condensed. I have analyzed the chordal conversions or transformations to be sometimes rather lengthy in their findings. Nevertheless, this is quite fascinating. It may be time-consuming to do these configurations; however, it is most interesting to find even the distant chord structures that are still related to the main tonic as if it was the same "bloodline."

    The examples I will offer at this point show that I have gathered some types of transmogrified chordal events; namely, in C Major: VI (E Major) of ii (G# minor) of V (F# Major) of ii (B minor) of V (A Major) of V (D Major) of iv (G minor) of ii (D minor) of I (C Major). Therefore, this is an octonary chordal transformation. Why go through this? My reasoning for this is that it is very interesting and, while it may take a lot of time figuring out, I am fascinated by the progressions and how to manipulate them by way of calculating them carefully. If there is any reason to alter or modify any of the chords, then it would be because there would have been some alteration required due to perhaps the extraneous nature of a certain chord or chords that may be difficult to ascertain and, therefore, may need to have one or more tones altered for that reason.

Chapter Two

Analysis of Movement I of Symphony No. 40 in G minor, measures 1-42, K.550 by W.A. Mozart

With reference to the first theme, the Tonic G minor is indicated for the antecedent phrase of the first five bars. The bass tone G is retained through measures 5 and 6 while the supertonic is superimposed over the G.

The first inversion V^6_5 of i is present in measure 7, leading to the V7 of i in measure 8. G minor is reverted back in measure 9, giving closure to the consequential phrase. In measure 10, we find the subtonic in its second inversion $Vii°^6_{4\ 3}$ of i because of the whole note C in the bass, measure 11 contains the i^6_3. Measures 12 and 13 have the same chords as measures 10 and 11 respectively. Measure 14 consists of an implied Vii°7+ (this is slightly altered) of iv of iv of i (Therefore, a tertiary chordal transformation) with an accented neighbor D in the violins. There is a German 6th chord in measure 15 followed by an Italian 6th chord on the last beat. The main functioning chord for this measure could almost be an implied V$_{43}^{6}$ of V of i as it is modified, except for the fact that there is, of course, an Eb in the bass. One might think of the chord on the third beat as a vii°$_{43}^{6\#}$ (modified) of iv of iv of iv of i. The last beat of measure 15 seems to suggest V7 of VI of iv of i. (The Bb in the implied V7 is not present and Db is enharmonically altered to C# in order to create the proper notation for the Italian 6th chord mentioned before.) There is an interesting interplay of the subtonic vii°$_{5b}^{6}$ (modified) of V resolving to that same chord V (D Major) of i in measures 16-20 involving a series of accented neighbor notes, C# being the main accented neighbor note among other accented neighbor notes in the woodwinds. The Ds in the strings function as a pedal point while the accented neighbor notes create a sense of temporary tension to each of their quick resolutions. With the first theme returned beginning in measure 20, we revert back to G minor in measure 22; thereafter, we find the iv$_{5}^{6}$ chord in measure 24 with its direction toward the root iv7 of i in measure 25. This iv7 of i may also be thought of as $V{}^{7}_{5\atop 3b}$ of V of V of VI of i. In measures 26-27, there is a V7 of V of VI of i, thereby creating a tertiary chordal transformation before measure 28. The following measures contain their respective chord structures: Episodic material occurs from measures 28 to 42: Measure 28: V (Bb Major) of VI of i. Measure 29: V^6_3 of V of VI of i; measure 30: VI^6_3 of i; measure 31: V^6_4 of VI of i; measure 32: IV^6_3 of i; measure 33: V^6_3 of VI of i; measure 34: V7 of V of V of VI of i; measures 35-37: the interplay of the neighboring chords V7 of V of V of VI of i to $ii^6_{4\atop 3}$ of ii of VI of i with accented neighbor notes again; measure 38: V to V7 of iv of ii of VI of i; measure 39: iv^6_4 of ii of VI of i to $vii°{}^{6}_{5b}$ of V of iv of ii of VI of i with pedal point F in the bass which may justify the clashing of the $vii°{}^{6}_{5b}$ chord which is superimposed over the pedal point F. Measures 40-41 have the same chordal events as in measures 38-39 with the resolution to F Major (V of V of VI of i) in measure 42. There is one complete measure of rest after this measure, which prepares for the second theme thereafter, which is in the key of Bb Major: again, V of VI of i. Notice I did not tonicize any of the chords other than the tonic itself. I went through all of the necessary permutations and calculated very carefully all of the chordal transformations.

Mozart Symphony No. 40 in G minor K.550

Analysis of Movement I of Symphony No. 40 in G minor, measures 1-42, K. 550 by W.A. Mozart

11

Chapter Three

Analysis of movement I, measures 1-66 from Symphony No. 5 in C minor, Op. 67 by L. van Beethoven

The opening generative cell of the first movement of Beethoven's Symphony No. 5 in C minor paved the way for one of the most memorable and appealing themes in the history of music. It undoubtedly originated from a bird singing those notes. We could imagine two phantom chords that would seem to be underlying for the two generative cells: those being i (C minor) followed by V (G7). C minor is present after the first 5 measures of introduction with the generative cell reiterated in the strings from measures 6-10. V^6_5 is utilized in measures 11-14 with similar cellular reiteration, except that the intervals have been either expanded or contracted using the same paeonic IVth rhythmic figure (... _). (Please refer to chapter 5 for details regarding musical scansion and its terminology.) Measures 15-19: i to V_5^6 i to V_5^6 to i. Measure 20: Contains an Italian 6 chord (this would have been a German 6 chord had Eb been included); measure 21: V; measures 22-24: iv (F minor); measure 25: still may be conceived in F minor although measures 26-27 with vii° seem to anticipate the movement toward V7 in measure 28. Measures 29-33 still have the same chordal interplay found in measures 25-29. In measures 33-37, accented neighbor-note motions pave the way for an aggrandized double neighbor note motion in the violins over the progression of ii°7 to vii$°_5^6$ in its resolutional state; iii+7 to i$_3^6$ in its resolutional state; superimposed over the C pedal point in the bassline. Measures 34-37, one finds the series of 7 to 6 suspensions along with the aggrandized suspensions to resolutions. Thereafter, in measures 38-43: one finds the continuation of this series of 7 to 6 suspensions to resolutions over the C pedal point in more diminutive note values: iv7 to ii$°_3^6$ of i; ii°7 of iv of i to vii$°_3^6$ of iv of i; iii+7 (no E♮ though) of iv of i to iv$_3^6$ of i; viiø7 (the seventh has been raised a half step to become Ab and the scale degree of A# has been enharmonically altered to Bb) of iii of V of i

Analysis of movement I, measures 1-66 from Symphony No. 5 in C minor, Op. 67 by L. van Beethoven

to vii°$^6_{43}$ of iv of i; iv$^7_4$ to iv$^6_4$ (iv7 to 6) of i; ii°7 to vii°$^6_5$ of i; measures 44-47: i (C minor); measures 48-51: V7 (G7) of i; measure 51: upbeat is in i (C minor); measures 52-56: vii°$^6_{5b}$ of V of V of VI of i. Or, if one were to perceive it another way analytically, it may be vii°$^6_{5b}$ of iv of iv of i. The problem with the latter configured progression is that moving from vii°$^6_{5b}$ of iv of iv of i to iv (Bb minor) of iv of i doesn't seem to be in proper harmonic orientation with the upcoming chord of Bb Major, although the nature of leading via the nature of the subtonic in this case would be a half step difference (C-Db) as opposed to the previous configuration which would otherwise be a whole step from C to D. The progression from C to D makes more sense. Therefore, a necessary alteration of vii°$^6_{5b}$ had been indicated. Measure 57 is kind of suspenseful in that there is a dramatic silence until the next chord in measure 58 of Bb Major (V$^6_3$ of V of VI of i) where there is another measure of suspenseful, dramatic silence. Thereafter, the intervallically expanded generative cell with its same paeonic IVth rhythmic figure begins in measure 59 with the French horn part having a strong announcement in fortissimo, bringing the harmonization to the relative Major chord of Eb, where the inversion of the downward perfect fifth in the French horn part is inverted to an ascending perfect fourth in the first violin part beginning in measure 63, creating the beginning of the second theme. There may be some controversy with regard to the paeonic IVth rhythmic figure in measures 59 to 60 as that rhythmic figure may become embedded into measures 61 to 62. Why? Because one could perceive the three eighths in measure 59 to be a tribrachys rhythmic figure followed by an anapestic rhythmic figure: uu− based upon the temporal-spatial plane. However, there may be some ambiguity with regard to this latter analysis.

Analysis of movement I, measures 1-66 from Symphony No. 5 in C minor, Op. 67 by L. van Beethoven

Analysis of movement I, measures 1-66 from Symphony No. 5 in C minor, Op. 67
by L. van Beethoven

Chapter Four

Patschen with Mixed Rhythms

*P*atschen is a German word meaning to clap. However, it also involves various forms of body percussion whether it's clapping the hands, slapping the knees, patting the thighs, or snapping the fingers. If one wants an efficient formula for performing different rhythms together at the same time, patschen is a way to make it more facilitated. For example, the left hand pats the left thigh in duplets. Once this is established, the right hand pats the right thigh with triplets. The initial beat is started with the hands patting exactly together at the same time. T=together. L=left. R=right. The formula for this pattern is: T R L R if the left hand pats the duplets and the right hand pats the triplets. This may be reversed in which the formula is T L R L as the reversal of the respective hand pattings is to be accomplished. The patting must be done with precision and evenness, and the beats of the patterns become equidistant in their execution. The formula for the triplets for the left-hand patting versus the quadruplets for the right-hand patting is T R L R L R. The reverse formula is T L R L R L, as the reversal of the patterns for the respective hand pattings is accomplished.

Patschen is one efficacious method to accomplish the beat patterns. However, it is not the only way. If one can figure how to break down various rhythms and subdivide the number of beats within the pattern against another pattern, much has then been accomplished. To show how five eighths can be set against two quarters, draw two sixteenths per each of the five eighths. Then draw two thirty-seconds per each of the sixteenths. If you want four eighths against 5 eighths, there will be five thirty-seconds (subdivided) for each of the four eighths. To do the patschen for a duplet with the left-hand patting against a quintuplet with the right-hand patting, try the following rhythmic pattern: T R RL R R with the central RL double-timed (in other words: RL twice as rapid). Reverse the pattern: T L LR L L (LR is twice as rapid). Notice that I did not space the central letters deliberately to show that they are double-timed.

If one wanted to try patschen with two eighths against seven sixteenths, draw two thirty-seconds per each of the seven sixteenths to show the subdivision as proof for mathematical accuracy. The first of the two eighths will obviously be with the first sixteenth. The second of the two eighths will be in between the fourth and fifth sixteenth, which is after the first set of seven thirty seconds that ends on the fourth sixteenth and the second set of seven thirty seconds begins between the fourth and fifth sixteenths. Now, to beat the two eighths against the seven sixteenths, try the patschen pattern with the eighths patted with the left hand and the seven sixteenths patted with the right hand: T R R RL R R R. (RL in the center of the pattern is double-timed: twice as fast). Reverse the pattern: T L L LR L L L. (LR is double-timed).

To enjoy more challenging rhythmic patterns, try various patterns to experiment with and have others try the patschen with slower or faster rhythmic patterns against yours, and work with simple and compound meters as well as odd-numbered patterns such as triplets, quintuplets, septuplets, etc. The results will be amazing as you perform the varieties, and feel free to utilize percussion instruments.

Chapter Five

Musical Scansion

Scansion is basically the scanning and marking of the stressed and unstressed syllables known as metrical or rhythmic feet, which are the units of measurement found in lines of verses, particularly in poetry with often a particular metric foot or a combination of feet. These are, of course, found in prose and various other forms of literature and in different languages. One finds the accented or unaccented rhythmic patterns to be related to the study of prosody, which involves the study of the elements of speech such as emphasized or de-emphasized rhythmic combinations of metric feet, pitch, intonation, tone, inflection, and duration. My point is that these emphasized or de-emphasized rhythms are found in music and many of the different forms of scansion are applicable to the temporal-spatial plane of music as well as the accentual-unaccentual plane. Herein lies some of the controversy, as several modifications may occur with regard to these realms. The macron (–) for a long or emphasized sound will have a durational alteration as it may be lengthened to a certain degree. If the macron does show emphasis, it will often be on a strong or semi-strong beat. It will at times even occur on a weak beat. If there are any exceptions so as to "go against the grain" so to speak, there may be some ambiguity regarding this phenomenon. The breve (u) for a short or de-emphasized sound would usually be on a weaker beat. Again, there may be some exceptions as the breve could also "go against the grain," and fall on a strong beat. The breve may also be modified regarding duration and de-emphasis.

The following list includes many of the metric-rhythmic patterns of scansion:

Macron: –	Breve: u
Trochee:	– u
Iambe:	u –
Pyrrichius:	uu
Bi-pyrrichius:	uuuu
Tribrachys:	uuu
Anapest:	uu –
Amphibrachys:	u – u
Dactylus:	– uu
Spondee:	– –
(or, Spondeus)	
Bacchius:	u – –
Creticus:	– u –
Antibacchius:	– – u
Paeon I:	– uuu
Paeon II:	u – uu
Paeon III:	uu – u
Paeon IV:	uuu – (This is a dead giveaway: Beethoven's opening generative cell of Symphony No. 5)

Epitritos I: u – – –
Epitritos II: – u – –
Epitritos III: – – u –
Epitritos IV: – – – u
Diiambe: u – u – (Two Iambic rhythms)
Ditrochee: – u – u (Two Trochees)
Ionicus a Majore: – – u u (Or, one Spondee + one Pyrrich)
Ionicus a minore: u u – – (Or, one Pyrrich + one Spondee)
Antispastos: u – – u
Chorjambe: – u u – (This is equivalent to a Trochee
  + an Iambic rhythm: Choree=Trochee) (Jambe=Iambe)
Molassus: – – –   (Just think, three molassus
  macrons could be as "slow as molassus")
Dispondeus: – – – – (Two Spondaic rhythms: or, two
  Spondees)

Chapter Six

Examples of Musical Scansion

Example 1
O Fortuna from Carmina Burana
by Carl Orff

Measures 1-3: The soprano, alto tenor and bass parts seem to have the paeonic I rhythmic figure. The temporal-spatial nature of the tones with the whole notes followed by the three half-note chords portray a kind of "head-tail" figure, whereby the first whole notes are elongated and the half-note chords are less long, and therefore one feels a sense of the paeonic I rhythmic figure, – uuu, the accents of which emphasize each of the chords with great strength, notwithstanding the last three de-emphasized breves based upon the main rhythmic scansion. Those three breves nevertheless still become emphasized with most likely an interpreted martellato.

Measure 3 contains a poco stringendo which enables the paeonic I rhythmic figure to be somewhat compressed, becoming rather closer to the true form of the paeon I itself, although still intensified. Measure 4 contains an anapestic rhythm: uu − In measures 5-8, the text, "semper crescis…aut decrescis" is seemingly congruent with its bi-pyrrichius rhythmic figure: uuuu. Accentuations go against the grain with regard to the Latin syllabication concomitantly with the placement of such syllables on either strong or weak beats. This makes the execution of the Latin text intriguing as the effect of the moon "waxes and wanes" and the vocal as well as the textual interpretation may follow suit within the general dynamic range of pianissimo. In measures 9-12: there is an overlapping across the barline with a tribrachys rhythmic figure: uuu, followed by the epitritos III rhythmic figure: − − u − with the words vita detestabilis (detestable life). The emphasized and de-emphasized syllabication again goes very much against the grain here versus the nature of the strong and weak beats. Interpretively, the impact of certain syllables will most likely be necessary to be deliberately emphasized: even though they are aligned with their opposing strengths or weaknesses as the case may occur. On the temporal-spatial level, there is an obvious, durational modification of the epitritos III rhythmic figure as the last syllable of detestabilis is elongated.

Example 2
Polonaise-Fantaisie in Ab Major, Op. 61
by Frederic Chopin

Before analyzing the musical scansion of the opening chords, the key signature is interesting in that Chopin hastily placed the third flat on the second line of the treble clef. Obviously, in error.

In analyzing the opening two chords, the first Ab minor chord with its upper sixteenth note (Ab Major: ii/IV/IV of I) moving to the second chord Cb Major (IV/IV/IV of I) with an intervening dotted eighth note; namely Eb, appears to be a bacchius rhythmic figure: u – –. Again, there is a durational modification of the second chord as it is elongated. One of the salient rhythmic features of the polonaise rhythm is an anapestic rhythmic figure: uu –. Chopin decided to add to that rhythm with two diminutive staccato sixteenths followed by the three staccato eighths. However, I think the rhythm translates to the following with some temporal-spatial modifications: uu (diminutive sixteenths) uu –, (uu uu –; uu –; uu uuuu)…etc. We have, in this instance, diminutive versus augmentative rhythmic patterns. I don't think it would be correct to indicate after the first two diminutive sixteenths two macrons suggesting a kind of molassus rhythmic figure: – – –, the third macron of which would be longer in value and not staccato but accented as Chopin had indicated. Therefore, one may desire to take under consideration a comparatively rhythmic differential processing of the sixteenths and eighths.

Example 3.
Variations on a Theme of Robert Schumann, Op. 20
by Clara Schumann

This has some very interesting rhythmic patterns. Measure 1: Because the first two quarters in the upper melodic line are portamento, I think of this as a pyrrichius rhythmic figure: uu followed by a trochee – u in measure 2. In measure 3, the quarters in the upper melody are not portamento but part of the legato line, which begins in the first measure and continues to measure 4. Because the quarters in measure 3 are connected, I would regard the rhythmic figure as a spondee: – – followed by perhaps another trochee – u in measure 4 for symmetrical purposes. Except, even though that creates consistency, there is a grace note at the end of measure 3 which anticipates the chord on the downbeat of measure 4. The upper dotted quarter followed by an eighth in measure 4 would ordinarily be a trochee. However, with the added grace note, it tends to belong to the rhythmic figure of measure 4. Therefore, is it not an amphibrachys rhythmic figure: u – u? Similarly, measures 5-7 are the same regarding the rhythmic patterns. However, at the end of measure 7, there happens to also be a grace note anticipating the half-note chord in measure 8. An iambic rhythmic figure u – perhaps? Or, if one disregarded the grace note, the half-note chord could be a macron that would sound alone by itself, still completing the phrase from measures 5-8. This architectural design or period of eight measures which consists of the antecedent (head) and consequential (tail) phrases is the beginning of the theme for which there are seven variations. With measures 9-16, the following rhythmic patterns occur: Measure 9: the antibacchius rhythmic figure: − − u; measure 10: the Trochee: − u; measures 11-12: antibacchius, then a trochee; measures 13-14: antibacchius, then an amphibrachys: u − u; measures 15-16: pyrrichius, then a macron − as a half-note chord sounding on its own in its resolutional state. Measures 17-18: pyrrichius, then a trochee; measures 19-20: dactylus: − uu; then a trochee; measures 21-22: pyrrichius, then a trochee; measures 23-24: dactylus, then a macron sounding alone in its resolutional state as a quarter-note chord.

Example 4
Prelude from Tristan and Isolde
by Richard Wagner

    Before analyzing with musical scansion, I first want to mention that there is a lot of chromaticism in this work. There is a great deal of tension with its suspensefully suspending chords; namely, the "Tristan" chord in measure 2, what would be enharmonically altered to a half-diminished chord followed by a French $^6_4{}_3$ chord, thereafter, an accented neighbor note in the upper leitmotif to V7 of i (A minor). I think that the downbeats of measures 1-3 suggest appoggiaturas as they all appear to be accented neighbor notes. They are elongated in the first two measures, the last of which is less long, yet still consists of heightened tension.

    With regard to the rhythmic patterns, the upbeat is much a part of the first rhythmic pattern up to the sixth beat of measure 1, which is seemingly an amphibrachys rhythmic figure: u – u. In measure 2, a trochee seems rather evident and in measure 3, an iambic rhythm appears u –. Why is the last figure not a trochee? Some may be of the opinion that because it begins with an accent, it's like the trochee based upon the accentual-unaccentual plane. However, I believe that the temporal-spatial plane takes precedence here and so, durationally, the accented neighbor, or appoggiatura appears to be iambic in nature. The quarter note B tied over to an eighth elongates the value of the B but it is not accented.

Examples of Musical Scansion

Example 5
Impromptu in Bb Major, Op. 142, No.3, measures 1-18
(the measure numbers are not including the repeat)
by Franz Schubert

    The theme of this Impromptu, which is an andante with five variations, begins with a dactylus rhythmic figure in measure 1. The second half of the measure has a pyrrichius rhythmic figure. Measure 2 also begins with a dactylus rhythmic figure and the second half of the measure has a half-note macron preceded by a grace note, thereby implying that it could be thought of as an iambic rhythmic figure. Measure 3 starts with a dactylus rhythmic figure again, the second half appears to be an anapestic rhythmic figure: uu –. Measure 4 begins with another dactylus rhythmic figure followed by a bi-pyrrichius rhythmic figure leading to a macron ending this portion of the theme. Measures 1-4 form the antecedent phrase. Measures 5 and 6 contain the same rhythmic figures as in measures 1-2. Measure 7 consists of the dactylus rhythmic figure on the first half of the measure, followed by a pyrrichius rhythmic figure. In measure 8, the dotted quarter note C in the theme is followed by an eighth note D, forming a trochee, thereafter leading to a resolving half-note macron Bb, giving closure to the consequential phrase of measures 5-8, giving closure to the first section of the theme.

In measure 9, beginning the second section of the theme, the dactylus rhythmic figure is present in the first half of the measure, followed by a pyrrichius rhythmic figure. Measure 10 has a dactylus rhythmic figure functioning as an upper neighbor note embellishment to the pyrrichius rhythmic figure on beat two. The second half of measure 10 has a trochee figure with an accent on the chord on beat three. Measure 11 has a dactylus rhythmic figure, then a pyrrichius rhythmic figure. Measure 12, again, has a dactylus rhythmic figure, then a pyrrichius rhythmic figure. Measure 13 has the same rhythmic pattern as in measure 5. Measure 14 is a bit more elaborate with its embellishments of upper neighbor and lower neighbor notes. The dactylus rhythmic figure is on the first beat. Without the grace notes in the middle of beat two, one would think the bi-pyrrichius rhythmic figure would be rendered. However, the added grace notes create an upper neighbor note motion in a diminutive rhythm, thereby enhancing the second beat with first a pyrrichius rhythmic figure at the beginning of the beat, followed by a diminutive pyrrichius rhythmic figure which consists of the two grace notes rounded out by the descending whole step as another non-diminutive pyrrichius rhythmic figure, resolving with a sustained half note macron. The same rhythmic pattern of the dactylus figure followed by a pyrrichius figure is found in measure 15. In measure 16, a trochee is utilized in the theme with its dotted quarter leaping up to an eighth-note F, resolving to a half-note macron. The same rhythmic pattern of measure 16 is utilized in measures 17-18 to conclude the second section of the theme.

Example 6
Verklärte Nacht, Op. 4 (Transfigured Night)
By Arnold Schönberg

This work, which is mostly considered a more tonal work, begins with a reiterating bass tone on D serving as a pedal point. This spondee or spondaic rhythmic figure – – is reiterated in the same manner up to measure 8. The main salient rhythmic feature seems to be the bacchius rhythmic figure throughout much of the beginning in the violas and violoncellos. The violins enter with the same bacchius rhythmic figure as well as an imitation of the theme along with various tonal harmonies that are incorporated throughout this magnificently structured work.

Example 7
Opening of the First Movement; Opening of the Second Movement from the Symphony in E minor, Op. 27 by S. Rachmaninoff

In the opening introduction, one finds upper and lower neighbor note figures in the violoncellos and doublebasses. The main salient rhythmic feature here seems to also be the bacchius rhythmic figure. The paeonic fourth figure, with its lower neighbor note motion, rounds out the end of the opening phrase. The woodwinds enter in measure 3 with an epitritos III rhythmic figure – not that the fourth beat of measure 4 is by any means short; however, it is comparatively less long than the other rhythmic components of this epitritos III rhythmic figure. When the violins enter with the upbeat to measure 5, we have a series of rhythmic figures; namely, a pyrrichius figure on that upbeat, afterwhich the series of bi-pyrrichius figures occurs. Interestingly, we almost hear the pulsation of the anapestic figure which is so much a salient rhythmic feature in much of Rachmaninoff's music and recurs quite often in this symphony. It is a kind of recurring trademark of his music that became a segment of the complete rhythmic figure, which is actually Rachmaninoff's signature rhythmically speaking: the syllabication of his last name seems befitting with the chorjambe (choree=trochee; jambe=iambe) which is the equivalent to the trochee or trochaic rhythmic figure + iambic rhythmic figure – uu – : Rach–man- in-off.

In the beginning of the second movement, one finds that anapestic rhythmic figure in the violins again and it also recurs in various sections throughout the second movement. In the third measure, there is apparently a paeonic IVth rhythmic figure followed by a series of iambic figures. Remember, there would ordinarily be a chorjambic rhythmic figure if not for the deliberate deletion of the first macron so as to create the anapestic rhythmic figure so prevalent in Rachmaninoff's music.

Examples of Musical Scansion

Example 8
Danse Russe (Trepak) from The Nutcracker Suite,
Op. 71a, measures 1-7
by Pyotr Tchaikovsky

In measure 1, there is a situation whereby the first chord is staccato. The first beat appears to be a dactylus rhythmic figure. Except, it is not compatible with the temporal-spatial plane. It is seemingly compatible with the accentual-unaccentual plane in that the first eighth is in fact short as well as accented, giving credence to the nature of the realm of emphasis. The realm of de-emphasis is obviously on the second beat and it contains the pyrrichius rhythmic figure with unaccented staccato eighth notes. The upbeat to measure 5, on the other hand, prepares a series of anapestic rhythmic figures: the same being true for the eighths, as they are within the realm of the accentual-unaccentual plane rather than the temporal-spatial plane. The macron preceded by the two breves is actually short, although less emphasized due to the fact that it is within a dynamically soft passage. This has nothing to do with the durational aspect of short-short-long but rather delving into the interpretive factor that undoubtedly would be influenced by some dynamic considerations, modifying the impact of the rhythmic figure to varying degrees.

Chapter Seven

An interview with the author, Philip Martorella, pianist, organist, composer, professor

Interviewer, John Faiola, director, actor, and vocalist

John: I am interviewing Philip Martorella, author of this fascinating book about musical theory and scansion in music. Welcome, Philip!

Philip: Thank you, John. I'm glad to be discussing my book and anything else that you desire to ask me.

John: I find your book very interesting. You do consider it controversial. Why is that?

Philip: I consider it controversial because there are various terms, expressions, and phrases that the reader might question and possibly disagree with, and I think that is perfectly understandable as well as respectable.

The reader has a right to his or her opinion and that is fine with me.

John: Suppose one disagrees with everything you wrote about...what if that happens?

Philip: I doubt that would be true. If there are such criticisms, I think what would happen is that I would respond with answers to determine why the reader would disagree and try to explain why I believe in what I have indicated in my book. Much of what I have indicated in this book has been examined carefully, researched meticulously, and I strongly believe that innovation with regard to a particular subject whether it be a discovery, a new invention, or some form of modification of existing ideas from one's accomplishments or insights from the past may develop into more improved ideas, and enhancements of any kind.

John: You obviously worked very diligently on your book. Hopefully it will be enjoyed by many readers. It is very interesting indeed.

Philip: I tried to make it rather interesting, of course.

John: Your background is very interesting. Your parents met at Lincoln High School in Brooklyn, New York and were married in Brooklyn. You were born in

An interview with the author

Brooklyn, and then your family moved to Long Island where you had much of your training. What was your age when you started to study the piano?

Philip: I was eight years of age. I first studied with my father, Sal, who gave me my first lessons, and I progressed rather rapidly through the graded books. I thoroughly enjoyed learning the music and I owe him a great debt of gratitude for what he taught me. He knew enough about the piano to teach me and my brother Stephen; and, my father wanted to give both of us the chance to have successful performing careers. My father was mainly a very fine violinist, and I believe my father's motive was for my brother and me to excel at the piano so that we would accompany his violin playing when we became more advanced. It was such a good idea to have such predetermined intentions in mind for his two sons.

John: Who was your teacher after you studied with your father?

Philip: My father then suggested a wonderful pianist and teacher, Avraham Sternklar, to whom I also owe a great debt of gratitude for his wonderful teaching. I have included a fascinating "teacher-pupil" lineage tree in the back of my book; and, because I had studied with him, the tree dates back to such composers as Beethoven and actually even further back in time to Haydn. There was that direct line to those great composers as well as other great composers and pianists on the teacher-pupil lineage tree and I'm proud to be a part of that tree.

John: You have a fascinating background and apparently come from a musical family. Your father was also an attorney at law during the day and during the evening, he would take his violin and perform in the orchestra.

Philip: My father loved to play the violin and he would practice in another room so as not to disturb me or my brother whenever we wanted to practice. We had two pianos in our living room and each of us would have our schedules as to whether it was my time to practice or my brother's time. We would strive to be fair to each other; however, that became challenging at times.

John: Your mother, Jeanne, was a fine dancer. She was not only a ballerina, but also a Rockette at the Radio City Music Hall. When was she at Radio City?

Philip: She was with the Rockettes during the 1940s under the direction of Russell Markert. She also had the opportunity to dance as one of the flower girls in the production of Bizet's Carmen at the Metropolitan Opera House in Lincoln Center, New York with Risë Stevens, soprano, who sang the role of Carmen.

John: When you were a teenager, you travelled with your parents to the Trapp Family Lodge in Stowe, Vermont, where you met Maria von Trapp herself. That must have been very exciting.

Philip: Yes, and I gave a piano recital there. When I looked in the gift shop, there was Maria at the registry selling various items. I noticed a beautiful old wooden flute. It was, of course, a recorder, much like the kind of recorders played by the Trapp family. I bought that recorder from Maria and I still treasure it to this day.

John: After studying privately with Avraham Sternklar for many years, it was then off to college, correct?

Philip: Yes. I went to Mannes College of Music and received four years of wonderful studies with great teachers of piano, theory, score reading, advanced score reading, analysis, composition, conducting, orchestration, etc. It was at Mannes College where Risë Stevens became President after Leopold Mannes and John

Goldmark served such a title. It was there that I studied with Nadia Reisenberg, a phenomenal pianist and teacher and her teacher-pupil lineage tree was also commendable. She had studied with the great pianist-inventor-composer Josef Hofmann, who, by the way, after noticing the movement of the metronome pendulum, actually invented the windshield wiper. He had a fascination with cars. He also constructed shock absorbers used in cars and airplanes and worked on piano actions for a certain kind of touch control. Nadia Reisenberg also studied with Alexander Lambert who was a pupil of the great pianist and composer Franz Liszt.

John: You are extremely talented and a very fine musician. This encompasses the gamut of musical expressions, like a palette of different colored paints. I understand that you also enjoy drawing and painting…

Philip: Yes. Drawing and painting are my other passions when I get the opportunity during my spare time to sit down and do some art work.

John: That's great. Getting back to music, you are an accomplished pianist. You also compose music and teach…

Philip: I enjoy practicing and rehearsing, whether it is solo or chamber music to prepare. I have a wide repertoire to choose from, and yes, I do compose a lot of music. I've written a piano concerto, several sonatas for piano, a work entitled Constellations for chorus and piano (four hands at one piano), two sonatas for viola and piano: one may be performed with clarinet; selections for piano such as etudes and several works for chamber ensembles.

John: I understand you performed the complete sonatas of Beethoven, the complete preludes and etudes, ballades, two sonatas, and other works of Chopin along with a vast repertoire of other music in many concerts. How do you find the time to rehearse all of this music?

Philip: It is a lot and I sometimes wonder myself if I am going to have enough time to rehearse everything that I set out to accomplish. Somehow, I seem to manage by figuring out daily schedules for rehearsing, teaching, composing, etc.

John: I must say that your background is very impressive. You have a Bachelor of Music degree from Mannes College, a Master's degree from the Juilliard School of Music, and you also received your teaching certificate at the Manhattan School of Music.

Philip: Yes, that is correct. I had excellent professors of music also at Juilliard.

John: You have a busy schedule currently. You resided in Queens after many years of study in Long Island. You gave a lot of concerts in New York. Then in 1991, you moved to the state of Rhode Island. I understand that you are on the faculty of music at Rhode Island College, correct?

Philip: Yes, and I'm enjoying that immensely.

John: I understand that you composed music for an album that you made several years ago?

Philip: Yes, the album is entitled "Relax and Dream," and the sales went extremely well. I received a lot of positive comments about my music for that album.

John: That's great. Now, getting back to your book, you really have delved deeply into the study of theory. It's quite impressive what you have written in your book. How long did it take you to write it?

Philip: Believe it or not, it did not take me very long at all. At least within a week.

An interview with the author

John: You recently went on a tour to Germany and Italy. Did you have the opportunity to perform in any concerts?

Philip: I actually played the organ at a chapel located in the Vatican, where the Italian Renaissance composer Giovanni Pierluigi da Palestrina conducted. I also had the opportunity to perform a piano concert In Italy. When I toured throughout Germany, I performed on numerous organs that Johann Sebastian Bach had performed on in various churches and cathedrals.

John: That really must have been quite an exciting experience. Hopefully, this book, along with the fascinating teacher-pupil lineage dating back to the great masters, will inspire readers interested in your experienced tutelage to want to study with you so as to keep such a phenomenal tradition. Keep up the great work, whether it's performing in the United States or abroad, or composing for solo instruments, combinations, orchestras, or being commissioned to compose for the theater, which I happen to know is in the works for some upcoming new musicals. Congratulations. You are a fabulous teacher and a dedicated mentor to many who want to continue their studies and would need your advice. What more could one ask for than to go to a master such as yourself. You are a consummate artist and you are very multi-talented. I want to wish you the best of luck with your career in music. Once again, congratulations, maestro!

Philip: Thank you, John!

John: Thank you, Philip!

Appendix A
Scores Referenced in Chapter Six

Scores Referenced in Chapter Six

Polonaise-Fantaisie in A♭ Major, Op. 61
by Frederic Chopin

Variations on a Theme of Robert Schumann, Op. 20 by Clara Schumann

Prelude from Tristan and Isolde
by Richard Wagner

Verklärte Nacht, Op. 4 (Transfigured Night)
By Arnold Schönberg

* Basses having no C string play only the upper notes.
P indicates a passage of primary importance.
S indicates a passage of secondary importance.
⌐ indicates the end of a passage of primary or secondary importance.

Scores Referenced in Chapter Six

Opening of the First Movement from the Symphony in E minor, Op. 27 by S. Rachmaninoff

Scores Referenced in Chapter Six

Appendix B

Teacher-pupil Lineage Tree

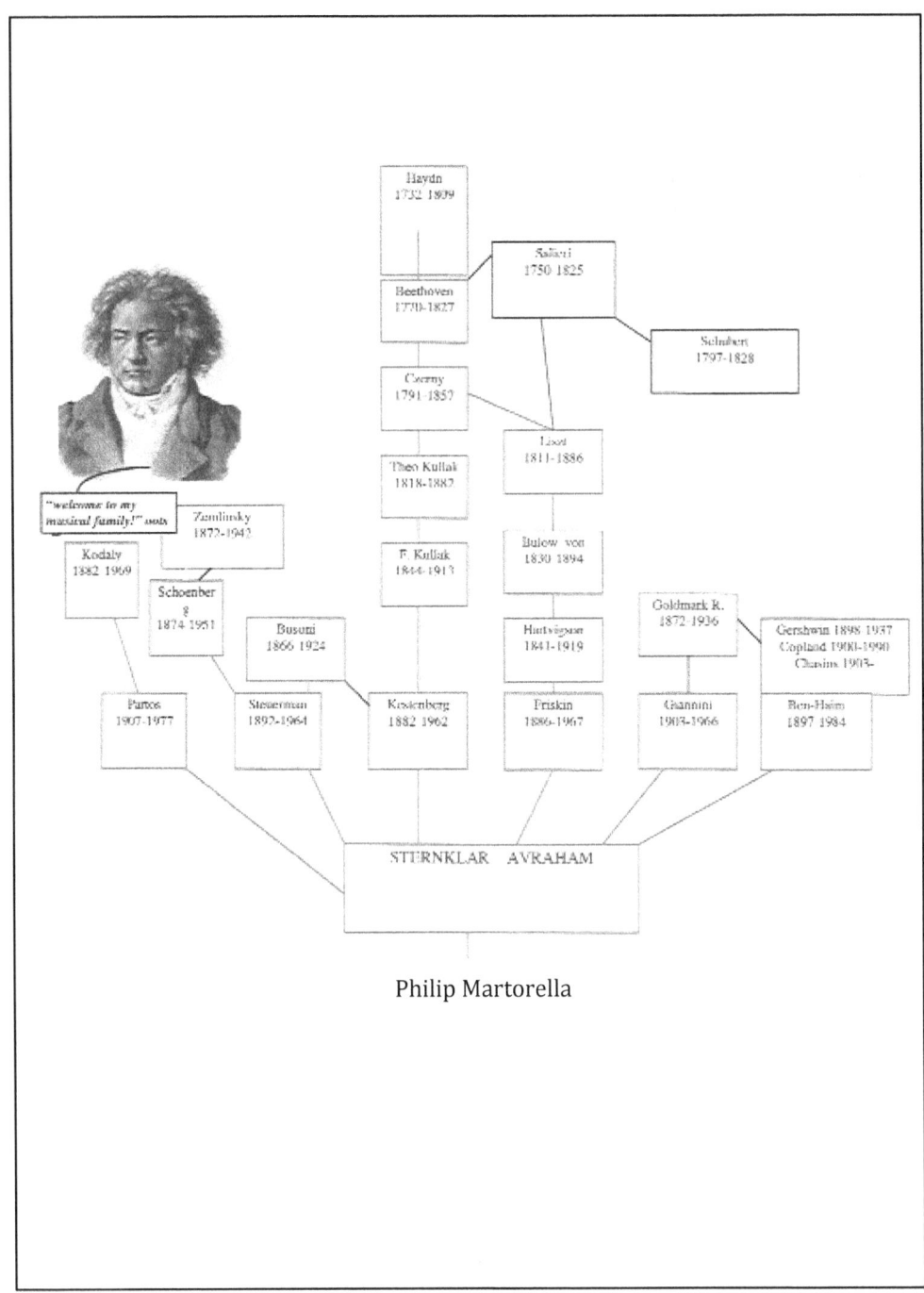

Philip Martorella

68

www.ingramcontent.com/pod-product-compliance
Lightning Source LLC
Chambersburg PA
CBHW072203160426
43197CB00012B/2502